Operational Reliability Manual

Author:

Eng. Alberto Windmüller

Original Title:
Operational Reliability Manual
Author:
Eng. Alberto Windmüller
Copyright ©2018 by Alberto Windmüller
First Edition
ISBN-13: 978-1723230455
ISBN-10: 1723230456

Every precaution was taken to assure the trustworthiness of its content; however, the author cannot assume the responsibility for the corrections that may arise from the information provided.

All rights are reserved. This publication may not be reproduced, in whole or in part, recorded in or transmitted by, a system of information retrieval, or by any means, be it digital, electronic, by photocopy, or otherwise without the prior permission from the author.

Index

Prologue	1
Introduction	4
Business Processes	7
Operation Process	8
Maintenance Operation	9
The Reliability and Integrity Management Process	17
The Process of Project Management	48
The Importance of Data, Quality of Data and Information.	59
Measurement of Process Performance	66
Epilogue	70
Bibliographic References	72

1. Prologue

This book is intended to serve as an introduction to Operational Reliability, although it also contains information that, being more detailed, serves to put the reader in the context that is necessary to better understand what is presented in it. Operational Reliability is a proposal for asset management that has served as a basis for developing many variations that have different names and certain additions, has not changed in its essence and has been reflected in an international standard that is ISO 55000 which is derived from a British Standards specification, the PAS 55 specification.

It is the author's conviction that it is very useful to understand which are the components that define what Operational Reliability is and how they interact with each other, contribute in a systemic way to the operational improvement of assets in an industrial facility. By capturing these concepts in a simple way in this book, we aim to ensure that the reader of any organizational level of a company is aware of what they are pursuing and how amalgamations produce a notable effect in increasing availability and efficiency in a productive system.

My experience in this discipline dates from the years 1988-1996. At that time the concepts, processes and methodologies of Operational Reliability were implanted in the Cardon Refinery of Maraven SA, a subsidiary of the company Petroleos de Venezuela. The company with the support of John Woodhouse and Jane Goodacre, consultants of TWP, initiated a project to transform the maintenance function to close the performance gaps that had been detected during a Benchmarking exercise that sought to measure the consequences that the implementation of a project to expand the refinery that would increase its complexity by 33%. By the way, as a result of the execution of this project, the other basic and support functions must have been transformed in order to adapt to the new demands.

With the passing of the years and assuming new positions, I had the opportunity to apply this knowledge and continuously improve it. I had the good fortune of working with people who increased my experiences either through the collaboration they offered me during our joint work or by offering me challenging opportunities that enriched me beyond Operational Reliability.

I want to thank the Creator first for giving me the immense joy of living and doing it in times that are extremely full of changes and challenges that made my existence interesting and full of opportunities to serve. I also have to thank my parents who sacrificed themselves so that I would receive all their love and support so that my education is the most complete, and this against all odds. They supported me in everything that is needed to grow in all areas of life. They taught me to be honest, persevering, compassionate and above all grateful. I also want to thank my extended family who has always accompanied me in good times and bad, but especially my wife Rosalba, who with her love and contagious energy has courted me and stimulated me on the path that leads me to be better every day . She is a vital piece for the publication of this book, which is the first. She surely will not rest in giving me support in the future. Many thanks also to my friends, co-workers and bosses who were a key factor in forging the knowledge I have poured into this book: Lino Carrillo, Ricardo Garcia, John Woodhouse, Jane Goodacre, Patrick Williams, Orlando Morean (especially his patience reviewing the manuscript) and Gustavo León. All of them contributed to making this book possible.

Surely there are many more that I should mention and that in the rush to publish and over the years I do not remember at this time. I can assure the reader that the comments and suggestions they have for improving this work are welcome and I promise to take them into account. In the future I will continue offering my experiences related to Operational Reliability, either concerning the associated methodologies, as well as the most delicate part that is the implantation because it has to do with the human factor.

2. Introduction

A home where everything works perfectly, where personal problems of each member of the family are analyzed and resolved by agreement, where honesty and integrity, the desire to contribute in favor of the family's success and development, and union are the pillars of their work on the medium and long-term, generates confidence.

This means that the family runs reliably for the benefit of the group and of each of its members, generating excellent results. This analogy can be used to introduce the concept of **Operational Reliability** in its simplest form.

What is Operational Reliability?

I believe it is a work philosophy, that having as basis total quality or continuous improvement, uses different methodologies that seek to keep operative the function of the assets (equipment, production systems, safety systems, etc.) according to the operating parameters for which they were designed.
However, this definition is extended with the introduction of other concepts that enhance it. The first is the notion of risk associated with the operation. When applying the risk criterion, priority is given to those functions where the risk associated with their failure or loss is more significant.

The other concept that contributes with the ampleness of operational reliability, is the life-cycle.

If all the factors contributing with the performance of an asset are considered in a project, and from its conception to the moment in which the industrial plant is disincorporated, the life-cycle of the asset is being discussed. Operational Reliability seeks to maintain the highest reliability of assets throughout their life-cycle.

Another concept that extends the Operational Reliability philosophy is the life-cycle cost, which translates into the consideration of all the costs that are generated from the conception of the project to the disincorporation of the asset or assets associated with it. When comparing different options within a project, considering the life-cycle cost of each one of them, the result is that the selected option has the highest probability of yielding the highest return on investment, provided that, the assets are managed in accordance with the Operational Reliability philosophy.

There is a diversity of methodologies associated with Operational Reliability, such as Reliability Centered Maintenance (RCM), the Risk Based Inspection (RBI), etc.

Other technical applications associated with the continuous monitoring of asset performance are also incorporated and will continue to be incorporated. Big Data, the Internet of Things and new trends in the application of artificial intelligence will substantially affect the management of Operational Reliability in the future.

It would be a substantial mistake if only the technical part is considered and the human and organizational components within what is Operational Reliability are neglected.

If you want to establish the philosophy of Operational Reliability in a Company, it is necessary to plan the implementation as a project considering especially the human factor. The success or failure of an implementation is linked more than anything to how change is managed during this process, and how the degree of enthusiasm of the company's organizations is enhanced to adopt and own Operational Reliability.

A component that is an important support during the implementation process, is the application of continuous improvement activities to establish new work processes or to improve existing ones, so that they adapt to the Operational Reliability philosophy and so that they support the creation of a culture where the capture, storage and analysis of process data is a vital part of the work of the operational units. This work must be on the top of the organizations priorities of the that interact in the routines of a facility, since it is the lever to operate excellently.

This book deals in short but in a very focused way with what Operational Reliability is and how it interrelates with the Maintenance and Project Management processes and shows how the continuous improvement and data management contribute to an excellent performance in the operation of an industrial plant or a larger facility.

During time, the concept of Operational Excellence has been stamped, which is nothing more than an expanded version of Operational Reliability.

3. **Business Processes**

The manufacture of bread with a commercial purpose obeys to established processes that even if they have not been modeled and developed in detail, should be considered as business processes that rule the commercial activity of a company, in this case the Bakery.

Production, maintenance of bakery equipment and infrastructure processes that are known to be universal are used for bread making. However, other less obvious ones are still important, such as the process of ensuring that the quality of the bread complies with sanitary norms and regulations, and that it is also an attractive product for the client. A significant number of processes associated with manufacturing companies are common to these activities and contribute to their success.

Correspondingly, to produce equipment manufactured in series, such as automobiles, washing machines, airplanes, etc. and continuous production companies such as oil and petrochemical companies, apply the same universal business processes.

For this book, we will focus on the business processes that have the greatest connection with value generation.

Figure 1 shows the Maintenance Process and two closely related ones that are, the Reliability and operational Integrity management process and the project execution process. These processes are intended to optimize the operations process, which is ultimately the added value mechanism of any company. In the following sections we will go into detail of all these processes.

4. The Operations Process

In a company, the operations process includes the set of processes, subprocesses, activities and resources used to generate added value and to satisfy customers with quality, productivity and profitability of the products and services offered.

In the context of this manual, the operations process will not be analyzed in depth because each company dedicated to produce goods or provide services has a different operating process.

However, there is a clear connection between this process and the Maintenance, Reliability/Integrity and Project Management processes.
The correlations with the Operations Process will be referred to in the discussion of the processes that we will describe next. The efficiency in the execution of these processes and their interrelations differentiate companies in their performance, where the best companies, place emphasis on seeking excellence.

5. The Maintenance Process

The Maintenance Process must be designed to execute actions in an efficient and effective way to ensure the reliability and integrity of the assets. Maintenance is ultimately aimed at preserving or restoring reliability and integrity.

The subprocesses related to the maintenance process are:

- **Detection of Maintenance Needs.**

    In this subprocess those activities of preservation and/or restoration of equipment reliability and integrity that need to be executed are selected, given their relative importance against the rest that are on the list of pending maintenance activities.
    These activities are materialized in the form of work orders (WO).

    The relative importance of maintenance activities depends on their nature, which in order of priority include:

    - ✓ WO for Preventive Maintenance
    - ✓ WO for Corrective Maintenance
    - ✓ WO for Emergency work

    The execution of an emergency WO prevails over of any other. This does not mean that the relative importance of the others changes. Only that the emergency work is executed immediately to ensure operational continuity.

The diligent and timely execution of the planned preventive and corrective WO leads to minimize the number of emergency WO.

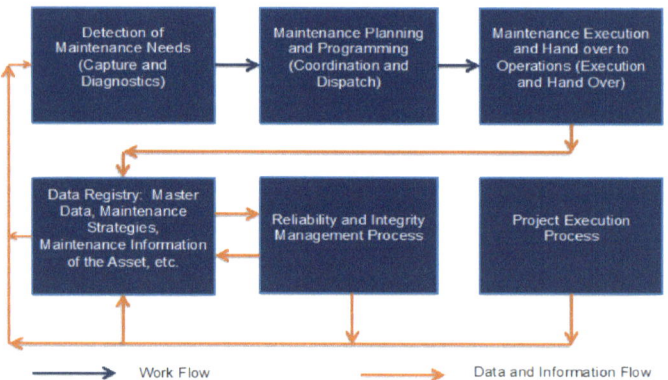

Figure 1: The Maintenance Process and its Connection with Reliability and Integrity and the Project Execution Processes.

- **Maintenance Planning and Scheduling**

    After selecting the WO for their execution, it is necessary to plan and schedule them in time.

    Good planning requires:

    ✓ A complete definition of the scope of work. This means that within this scope, all the work that must be executed to preserve or restore the assets reliability and integrity, as well as those secondary works necessary for the activities to be

executed safely and efficiently are included.
- ✓ The task list that must be carried out in sequence to execute the maintenance work according to the execution rules, as well as according to the quality requirements specified in the task list. This list should specify the tools and materials necessary to professionally complete the tasks.
- ✓ All the materials and spare parts necessary for the maintenance of the asset, well identified and according to the materials list and supplies that must accompany them. These should include special tools and measuring equipment and quality control if necessary.
- ✓ A checklist and delivery protocol to operations that must be completed during the hand over process. In it, during start up the measured operating parameters must be recorded.

NOTE: The WO must consider within the schedule, the necessary time to upload the information related to the completed activity including the operational values that the asset showed during the trial process in the hand over to operations.

- **Execution of the Maintenance and Hand over to Operations.**

The maintenance activity must be carried out by qualified personnel. An indispensable condition is that the personnel must be trained to perform the tasks safely. The following should be considered:

- ✓ Request the work permit and ensure that its scope corresponds to the nature of the maintenance activity that requires it.
- ✓ Confirm the isolation of all energy sources related to the equipment to be operated (electrical, pressure, temperature, etc.).
- ✓ Carry out a risk analysis at the job location level and identify external risks and eventual escape routes. In case of doubt consult with the operations personnel to clear any uncertain conditions.
- ✓ Prepare the equipment to be intervened, watching that it is in good cleaning conditions and not energized.
- ✓ Review the task list and become familiar with it. Confirm that the materials and tools are complete and in good condition.
- ✓ In the event of finding a condition different from that diagnosed in the equipment related to the work order, an analysis should be made to determine if there are additional risks involved. If the risks of the new condition are not duly covered in the plan, the intervention must be rescheduled to avoid unnecessary exposure to risks by the personnel.

Once the intervention is completed, the maintenance personnel must deliver the asset to the operator. The operation parameters of the equipment in the expected normal ranges must be checked and in the hand over protocol the conditions of housekeeping of the place will be included. The last step encourages a culture of care of the assets that ensures that housekeeping is an integral part of the job.

- **Data registration**

In all human activity, data and information are often neglected and these are a very important part of any productive task. Data and information are vital components for the continuous improvement and their collection and assurance must be encouraged and regimented.

Figure 1 shows a sub process that is part of the Maintenance activity and it is common to the processes of Reliability and Integrity and Project Execution. This is the sub process of Data Registration.

Figure 1 shows the flow of data and information of the different processes involved in operational reliability.
From the point of view of the asset life-cycle, the original data must come from the project execution process.

When a project is developed, one of the first activities that occurs after having defined the value generation process, (primary process of the operation), is to define the hierarchy of its components depending on the functions they perform and the relative importance that they have within the operation.

This essential step allows to relate these components with the Maintenance, Reliability and Integrity processes.

Together with the hierarchy the Master Data of the Asset must also be defined. These describe the most important characteristics and operational parameters of the assets. This master data is the asset's birth certificate.

To have an asset operating reliably it is necessary to ensure the existence of parts that are required for any repair or maintenance. It is therefore necessary to have a category of master data that relates the asset to its components, and parts. This category of master data is the so-called Material Master Data. Other materials also are part of the material master data, even when they are not related to the primary operations of the asset, e.g. consumables. These materials are necessary to ensure the operational continuity of the assets.

During the operation cycle of an equipment or asset, its performance deteriorates due to wear, undesired operating conditions, etc. When the performance of the asset is not within the minimum acceptable ranges, the asset must be temporarily disincorporated from the operation to restore its reliability /integrity through a process called major repair or overhaul.

The maintenance activity begins with a diagnosis of the assets condition, and it is very important to capture the data of its condition to determine the necessary repair measures to restore the asset for a lasting reliable operation.

The data on how the asset was found at the time when it entered into maintenance should be recorded and kept in the equipment's maintenance log, as it is important information to define future steps related to its performance, as well as to execute service life analysis.

While operating equipment and to ensure a performance in accordance with the established parameters, performance values are measured and recorded, which must be preserved for future analysis as well as to demonstrate compliance with government regulations.

Likewise, after having restored the reliability to an asset, it is necessary to collect data to be recorded in the equipment maintenance log for later use in case of failures or performance deviations.

During the discussion of the Reliability and Integrity process, we will delve into what it means to capture data and information that helps to continuously improve the asset performance. The management of data related to this process is an important aspect of the management of business assets.

Collecting, recording and preserving data and information on assets is an extremely important task in the management of a company and must be taken very seriously. These activities should be part of the Operational Reliability culture.

A small company that does not have enough incentive to install specialized systems for asset management, can use office tools such as Excel to develop their management support systems. However, there are products in the market of moderate cost that allow the development of an adequate database for asset management, with the ability to plan, schedule, keeping inventory of parts, etc., what is needed for professional management of the maintenance function, as well as operational reliability.

More sophisticated systems to support asset management allow much more complex activities and have as a competitive advantage an integration of various support modules that interact with each other. Others are even more sophisticated and allow interaction between asset management functions and support functions (Finance, Human Resources, etc.).

It is important to have a system that allows the storage, management and use of data in such a way as to allow the measurement and control of the maintenance function, Reliability and Integrity and of Projects that interact to offer control of the life cycle cost.

6. The Reliability and Integrity Management Process

The Operational Reliability Model

The processes that are being addressed in this book are those that interact in what refers to Operational Reliability.
Operational Reliability is a different approach to reliability that not only focuses on the inherent reliability of the asset or equipment, but also integrates the other processes related to the Operation process to connect them so that the result is optimal.

This model has four components that combine to produce a result that not only considers the reliability of the assets, but also introduces concepts that, when applied well, improve their performance within the operational context. It is a holistic approach designed to achieve quantum changes, as well as continuous improvement.

These concepts are: Human Reliability, Reliability of the Design, Asset Reliability, and Processes Reliability (Figure 2).

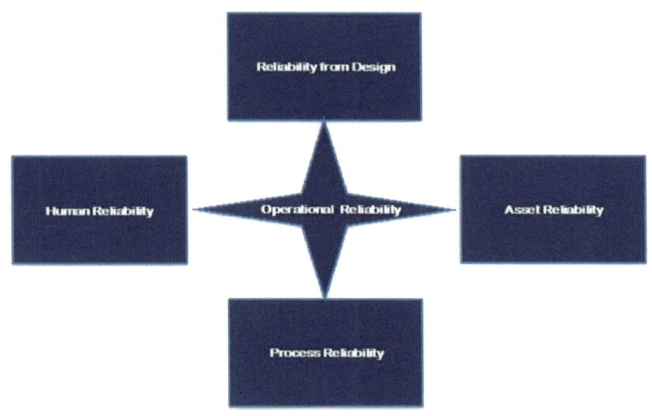

Figure 2: The Operational Reliability Model

Reliability can be applied to many value-generating activities. Manufactured products are often referred to as having high reliability and are required critical elements of a process or an equipment, which demand a high reliability, and is measured in terms of percentage.

The human being is the main actor of all productive activity. As such, it plays a key role in the implementation of what is Operational Reliability. Special importance should be placed on the management of human resources when Operational Reliability is being introduced in a company to guarantee a successful outcome. It is important to emphasize in obtaining and keeping a workforce that is prepared to support an excellent management of Operational Reliability.

A holistic approach related to Human Reliability is the key to success in any company that is venturing into the implementation of initiatives that are aimed at improving the performance of the company.

- **The Human Reliability**

   Human reliability refers to the probability that a group of people engaged in the processes of generating value in a company will be able to execute them without errors or failures, efficiently and effectively, within their scope of competence and in a specific organizational context.

   Human reliability depends on various aspects that provide an individual with the abilities required to execute the processes under his responsibility, so that the probability of human error is at the level of acceptance required for the tasks he performs.

   The aspects that promote competition and therefore contribute to having a high human reliability are the following:
   - ✓ Training of the individual
   - ✓ Sense of belonging
   - ✓ Motivation for Achievement
   - ✓ Communication ability

- ✓ Ergonomic Aspects

- ✓ Personal Development

- **Training of Individuals**

    Undoubtedly, Training Individuals is the most important subset of the aspects that combine to achieve high human reliability. It is my conviction that the efforts dedicated to the training of the employees of a company, provide a high return on investment.

    Training of human resources has different aspects and without a doubt the platform to have a competent workforce is having selected it appropriately, so that it performs in the aspect to which it will be dedicated to. This platform also includes the additional training necessary to cover knowledge gaps that cannot be expected to be part of the study/training curriculum taught by an educational institution. These gaps must be closed by means of formal instruction, mentoring and assignments that require a personal initiative to take on the challenge of fulfilling the ordered task. This last modality of learning sets the knowledge in a much deeper way and has as an additional benefit, that it contributes to the empowerment of other aspects of human reliability such as, for example, the consolidation of the individual's sense of belonging with respect to their environment. responsibility elevates motivation to achievement, among other things.

Having competent personnel not only means that they can fulfill the assigned tasks, but they must also have the awareness that being competent means seeking excellence. That is, they must provide all their knowledge, dedication, rigor, perseverance, the desire to improve the results of their work and be completely honest with themselves and with their environment, when analyzing the situations with which they may be confronted with. This is the only way to capitalize on the day by day learning that the interaction with a world where the complexity of work increases exponentially offers us, and competitiveness is a tool that must be continually sharpened.

To achieve that the human resources reach the level of competence described above requires from the supervisor, the manager, an important dedication of their effort (and part of this effort is time) to measure the skills of their personnel and ensure that there is a systemic growth of skills and capabilities, as well as those other aspects that lead to high human reliability.

Being competent also means having abilities that facilitate interaction in the workplace, where the speed to make the right decisions and put them into practice is increasingly vital to maintain a competitiveness that guarantees the feasibility and sustainable success of a company.

These abilities have to do with the soft part of the work environment. These are rather skills to interact successfully with a world where teamwork, collaboration between departments and the ability to innovate are tools of the daily work and require from the actors of this new way of working to be enabled for it. Companies must make an important and honest effort to encourage the development of these skills and empower employees to apply these skills without interventions that undermine them.

This way of working requires that managers acquire new skills to direct the workforce and for them to be mentors, guides and trainers of human resources, where the measurement of personal performance is a tool to achieve the results expected by companies.

- **Sense of Belonging**
  The sense of belonging is an ancestral feeling that has to do with the satisfaction of the desire to be part of something important or to possess something that satisfies personal needs.
  Its root is in the family, the group or human nucleus where belonging to these represents security, since it satisfies the needs of the individual that is part of it, whether it is of love, protection, sustenance or personal development. As a member of it, it does everything in its power to ensure its existence.

We can take this same concept to the business world. The company that creates an environment where the feeling of belonging to a group, that strives to generate value and that induces employees to feel ownership of their achievements and successes, will have a workforce that will respond much more effectively and efficiently to the challenges that may be imposed on them.

The employee with a high sense of belonging will make much more of an effort to generate value from their individuality, belonging to a team, a department or a work crew.

To create a favorable environment where the sense of belonging flourishes and extends, it is necessary to satisfy the basic needs of employees, such as the enjoyment of an adequate salary, offer them health insurance for them and their family, allow them to grow economically, etc., and also make sure that everyone has a clear understanding of the mission and vision of the company, its values and principles, the objectives of generating value that the organization has set for the period and most importantly, that it has clear objectives and personal measurable goals and actions linked to the objectives of generating value that the company has set.

Likewise, for the sense of ownership to become part of the organizational culture of the company, it is necessary to fight against any generator of fear, because fear is the worst enemy of a company that aims to promote excellence, since it discourages or annihilates continuous improvement.

The sense of belonging is an important element that contributes to the success of a culture of Operational Reliability and must be encouraged to implement and maintain it.

- **Motivation to Achievement**

A high-performance athlete gives his best to prepare, achieve competition and reach the highest-ranking positions in the world where he is compared. That desire to become the best, can be called Motivation to Achievement. He prepares himself conscientiously, invests enormous efforts in time and dedication to be able to demonstrate that he is the best.

In the corporate world it is necessary to create and maintain an environment where employees have a high motivation to achieve their goals and objectives. This environment should not be overwhelming, but rather attractive.

Measuring is the key word to promote motivation to achievement. If the goals and objectives of a worker are clearly associated with the highest objectives of the company and if they can be measured so that the effect they have on the achievement of higher purposes can be evaluated, a positive effect can be caused on this worker, since it can enjoy the success of having contributed to the creation of value.

A healthy compensation policy associated with performance and others, that aim to recognize the achievements of those who give their best effort promotes the Motivation to Achievement of employees and regarding Operational Reliability, measuring the availability of production facilities generates a motivation of collective achievement in those employees related with the facilities where they have a shared responsibility.

- **Communication Ability**

Communication is the action of transmitting information from a sending entity to a receiving entity. It is successful at the time when the transmitting entity confirms that the receiving entity has managed to interpret it faithfully according to its purpose.

Communication in any field of human activity is the lubricant that makes things flow. Well-conducted communication promotes the success of people and groups where activities that require coordination of resources are developed, clear instructions, feedback of information and finally any action that leads to a clear understanding of the purposes for which these activities were established.

The Communication Capacity is a characteristic that supports the level of success of a company.

Being able to communicate depends on several factors that make communication effective. The first is respect, that must be offered and received when communicating. The second is honesty that must prevail during the communication. The third is the level of freedom to communicate without fear that must exist in an organization for it to be effective. The fourth, refers to the company's emphasis on having employees that have the skills to communicate effectively verbally and in writing.

The Communication Ability in a company depends on how its leaders model and demand from the employees, regarding the facilitating factors of communication mentioned above. It also depends on the effort that a company puts into educating and training its workforce to achieve a high Communication Ability.

Without an adequate Communication Ability, Operational Reliability will be limited.

- **Ergonomic Aspects**

  According to the International Ergonomics Association, ergonomics is the set of applied scientific knowledge so that work, systems, products and environments are adapted to the physical and mental capabilities and limitations of the individual.

To be able to produce in a reliable way, it is necessary to feel comfortable and safe in the workplace.

If the work environment has conditions that are distracting, that generate fatigue or represent a risk factor for health, personal safety or the production process, human reliability deteriorates or is nullified. It is necessary to ensure that the ergonomics factors related to the job being performed have been analyzed and adequately conditioned so that the interrelation of the human being and the operational environment is adequate and does not result in incidents that endanger the safety of the individual, the reliability of the operations and the availability of the productive train.

Human reliability depends on the ergonomic conditions that prevail in the workplace and it is therefore of utmost importance to ensure that ergonomic aspects are considered from the inception of a project and planned for the life-cycle of the asset.

- **Personal Development**

For workers at any level of an organization to remain constantly motivated to achieve, their work must be interesting or better yet exciting. It is also necessary that aspects of ergonomics related to the workplace are optimal. More important yet is to show the employee what their future will be within the company.

To ensure job stability, good performance and personal satisfaction, the employee needs to have a clear vision of how they will grow professionally within the work environment.

Depending on aspects such as their potential, positions available for development, inclinations of the employee with respect to his career and level of performance, it is necessary to design a development plan where, in addition to the employee's professional progression, the requirements they must fulfill so that this plan can be executed must be specified. This development plan is a commitment between the employee and its supervisor that is reviewed annually to define the degree of compliance and to develop strategies to close any gap in the professional or leadership skills that may delay or impede professional growth. Like any plan, it is perfectible and must remain flexible to avoid decisions that generate dissatisfaction, uncertainty or disappointment that jeopardize job stability.

When an employee knows his professional progression, his degree of commitment is greater and distractions that could put human reliability at risk are avoided. The company's policies must be clear regarding personnel development and to ensure that there is a precise definition of their training.

- **Reliability in Design**

We have all gone through a situation where we regret not having done something earlier because we did not foresee negative consequences. Surely, we will realize if we make an analysis that the cause of this situation refers to that we saw things only from a point of view, without considering other factors that may eventually influence the result.

If we start to see how some infrastructure projects are handled, project managers only consider, in a narrow-minded way, the cost and the schedule. Aspects of quality, operability, maintainability fall outside the scope, and additionally no considerations are made for the divestiture of the asset at the end of its useful life. Even compliance with environmental regulations and safety process are rejected.

From the Operational Reliability point of view, a project must be conceived considering the life-cycle cost. The design considerations of the project must consider all the aspects mentioned in the previous paragraph. It is also advisable to imagine any future expansion or improvement, so that the mentioned aspects are analyzed to avoid surprises.

The aspects of operational reliability that have important repercussions on the generation of value that the asset should offer, are related more than anything to the maintainability and operability that are closely connected to reliability.

If a reliability approach is employed during the design of the asset, it is ensured that the projects will have a higher probability of delivering a facility that will show superior performance compared with others where the aspects indicated above were not considered.

Reliability in the Design Phase makes use of reliability methodologies to support an optimal cost-benefit throughout the life-cycle of the asset.

There are methodologies such as Reliability-Availability-Maintainability Modeling (RAM), which provide invaluable support related to the definition of an optimal configuration of an industrial plant. By this I mean that during the conception of the project decisions are being made that will impact the future performance of the asset considering the reliability aspects that will show the best cost-benefit in the long term. This methodology provides information related to the industrial plant's availability that can be modeled according to the initial process diagrams, so that in the earliest period of the project cycle, decisions can be made that optimize availability throughout the useful life of the asset, considering among others, aspects of production planning, industrial, environmental and safety regulations.

During these early stages, it is also important to consider the reliability and integrity aspects of the equipment, so that the cost involved in maintaining high personal and process safety is included in the life cycle cost. Aspects of Ergonomics are an important part of the analysis so that human reliability is included within the design.

During later stages of the project and before initiating the commissioning and delivery, the aspects of reliability/integrity, maintainability and operability should be reviewed in greater detail. Methodologies such as Reliability Centered Maintenance (RCM), Risk Based Inspection (RBI), aspects of Total Productive Maintenance, among others, will serve as support so that the assets can be maintained and operated to provide the best long-term performance. All the information that is developed during the application of these methodologies should be part of the documentation that the project delivers to operations and maintenance (part of the Deliverables) so that the asset can be started with a clear vision of the asset care strategies that the facility requires to meet the proposed availability goal.

Reliability activities that are developed in a project should be managed by reliability, maintenance and operations personnel assigned to the project so that they are responsible for implementing and applying the asset care strategies that will ensure the expected availability.

It should not be forgotten that the early application of methodologies of Process Safety Management such as HAZOP, HAZID, etc., also contribute to the goal of designing a reliable asset and at the same time ensuring the integrity in the operation of the asset.

- **Asset Reliability**

    One of the most commonly used definitions of Reliability states that "Reliability is the probability that an asset fulfills its function framed within a defined operational context".

    Within the context of this book, it will not delve into the already known concepts of Reliability. We will comment on the application of the principles that support the increase of asset reliability of a production center.

    The Risk criterion is fundamental to be able to efficiently manage Operational Reliability. There are several methods to define the risk associated with operational reliability. For the application of almost all the methodologies that support it, the use of the concept of risk determination is fundamental to gauge the relative importance of assets and events related to their condition.

    To improve the performance of a productive system, the first thing that must be done is to solve the problems that distract resources (financial, personnel, etc.) and the attention of management. Many of these problems not only affect asset availability but represent a high safety risk during operations and endanger the reputation of companies. Reacting to these problems can be very arduous with respect to its consequences, and the attention that the personnel must put on operations, maintenance, control of production processes, but more importantly, facilities availability and their effect on the generation of value.

It is necessary to solve these problems definitively so that they do not recur. For this it is necessary to develop a system that we can call "Defect Elimination Management System."

This system must have the information related to failures and defects of the assets, the risk they represent for reliability, personal and process safety and their consequence on value generation. To do this, a failure database that collects this information must be developed.

The Defect Elimination Management System for the of must be governed by a clear process and rules for the definition of the priority with which the problems must be approached, as well as the way in which the teams dedicated to its analysis are selected and the implementation of the recommendations that derive from the analysis. The concept of risk applied to this system helps to prioritize the problems, as well as the definition of the level of scrutiny necessary to obtain a result that covers the expectations of the personnel involved.

The applied methodology for the problem resolution that is most successful in the industrial, medical, managerial world, etc. is the so-called "Root Cause Analysis", applied by countless companies and institutions worldwide.

Of course, there are numerous variants available in the market. The principle is the same and the important thing is to understand the methodology, to apply it consistently and develop a group of people who come to have expertise in its application. These experts will be responsible for applying the methodology, in such a way that there is consistency in its use and thus ensuring a uniform quality in the results.

The rigor a company puts into the implementation of the system, the follow-up by the management of the execution progress of the actions proposed by the team responsible for the root cause analysis applied to a defined problem, and the development and monitoring of key performance indicators related to the system, will be the key to creating an operational reliability culture.

By solving the problems that constrain organizations, time and resources are released to undertake an improvement of reliability from the point of view of failure prevention and its associated costs (unplanned maintenance and deferred production). It also creates the opportunity for the personnel to be more innovative and productive.

One of the preventive methodologies that is used a lot to generate an increase in the reliability of the equipment is the Reliability Centered Maintenance (RCM). This methodology whose genesis goes back to the years 1961-1970, was developed by the aviation of the United States of America to increase the reliability of the components of its airplanes and at the same time reduce the maintenance intensity applied to them. The maintenance costs were very high and after reviewing the effects of this intense activity on the failure rate, they determined that there was no direct correlation to improvement. Many resources were being used and the improvement was minor.

A methodology was then developed that determines the different failure modes that the assets could present, hierarchizes them and defines the necessary activities to restore reliability so that the asset continues to fulfill its function or functions, concentrating the maintenance effort in those modes of failure that really require it.

This methodology showed that a lot of money is usually spent on the care of assets that could be used for other purposes of adding value. Companies that are already immersed in a preventive maintenance regime, if they apply RCM can reduce their spending by 40% -70%.

The application of RCM has gained great acceptance and has proven its enormous usefulness. However, when applying RCM, it is necessary that prior to this a hierarchical exercise (Criticality Analysis) of the assets of a facility must be done by applying methodologies to define the Criticality of each component of the industrial plant's systems, depending of its function to then determine which of these functions and their sub-functions require the RCM application. This is necessary because applying this methodology can be very onerous. If it is used without this previous analysis and with the rigor required to execute it properly, it would take a lot of man hours of maintenance personnel, operations, inspectors, process engineers, etc. The application of the RCM methodology demands rigorous facilitation, which is another factor that must be considered when deciding which equipment requires it.

There is an infinity of methodologies derived from the original RCM, which adapt to lower risk conditions and simplify the definition of maintenance strategies. It is necessary to analyze which one is appropriate and take advantage of the use of templates associated with the different types of industrial plant assets that can be modified depending on the criticality and design characteristics, to be more efficient in defining the maintenance strategies.

Steps for the development of a Maintenance Strategy

- Establish Context / Process Objectives
- Establish Maintenance Objectives for the Unit
- Establish Maintenance Goals for the Plants
- Establish the Maintenance Registry of the Unit / Plants
- Determine Function and Equipment Criticality
- Consider Maintenance Options
- Formulate Maintenance Strategies
- Generate and Authorize the Maintenance Quality Plan
- Effects on Spare Parts Policy
- Generate Detailed Work Instructions
- Implement Maintenance Quality Plan
- Gather Performance Data
- Continuous Improvement Processes
- Review the Criticality Assessment

Figure 3: Steps to develop a Maintenance Strategy

It is imperative to ensure that the personnel involved in these tasks have the necessary expertise and experience to guarantee quality. It is a good practice for personnel who have a sense of belonging and knowledge of the industrial plant to be dedicated to these tasks.

The operational reliability of an industrial plant also depends on the integrity of the pressure systems (Pressure vessels and piping) that are part of it. When talking about integrity, it is about the capacity of the pressure systems to retain the fluid they handle without leaking, that is to say, to hold its leaketightenss.

Depending on the pressures, temperatures and the capacity of the fluid handled to deteriorate the material of the pressure system, actions must be taken to avoid events that generate process interruptions, safety incidents (fire, explosion, environmental contamination, etc.) and consequently, asset stoppage with lost or deferred production.

The inspection of the pressure systems (pressure vessels, pipes, tanks, relief valves, etc.) is used as a tool to determine their level of remaining integrity and depending on the proximity to the limit values calculated and regulated by standards and design and process safety regulations, we proceed to depressurize the systems and take actions to restore integrity.

The traditional approach of the industry considered the inspection of the vast majority of pressure vessels and the total length of the pipes of a facility to then act on those assets that needed to be repaired to restore their integrity. Personnel with inspection experience could restrict the scope of inspections and consequent repairs based on their experience and the results of previous inspections of these assets. This way of defining the scope of inspection was extremely difficult and did not guarantee the reliable operation of the industrial plants, because there was always the danger that a bad operation that was not reported could accelerate deterioration of the pressure system.

Efforts led by the American Petroleum Institute (API) resulted in the development of an inspection methodology that, based on the risk of leakage of an asset, sets inspection intervals and defines the methods that should be used to inspect it.

This methodology is known as Risk Based Inspection (RBI) and is being applied around the world.

The Risk Based Inspection is a methodology that adheres to following steps to determine the frequency and the application of the different inspection methods available to execute asset inspections, depending on their level of risk within the productive process in which they are operating:

- Segregate the industrial plant by the different physical and/or chemical conditions that apply to the transformation and/or separation of the products it processes.

- Carry out a corrosion study of the group of assets corresponding to the conditions subject to analysis. The study of corrosion is a systematic analysis of the physical and chemical phenomena that occur in the process and the effect they exert on the materials that were selected for the construction of the plant, along its route.
- A flow chart in Figure 4 describes the process to follow.

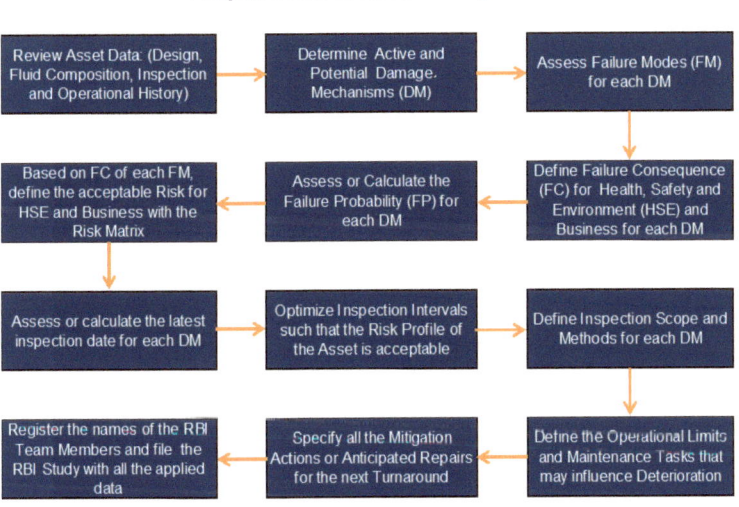

Figure 4: The Risk-Based Inspection Analysis Process

Afterwards the continuous improvement cycle is applied, and the process is repeated with the new data collected during the inspection at plant shutdowns. The results of the new run provide the new intervals (dates) and methods to be applied during the next inspection.

It can be said that there are other different methodologies that are aimed at improving or optimizing the reliability of assets. Some are condition monitoring applications that can be handled manually or electronically and that allow determining the optimal moment where the condition of the asset requires it to be removed from operation to restore its reliability.

A proper maintenance methodology that indirectly improves the reliability performance of an asset is the so-called Total Productive Maintenance (TPM), which essentially empowers and educates the owner of the asset that is operating it, to perform cleaning tasks, minor maintenance as lubrication, for example and monitoring and condition registration that allow operation within the operating window. It also allows the operator to do small analysis exercises with the data collected during monitoring. This information is valuable when it is necessary to practice a deeper analysis of the asset.

There are also statistical methods that support the analysts to determine the reliability of systems and individual assets. Weibull Analysis and various applications such as the Crow-AMMSA Reliability Growth Model are very useful for issuing behavior predictions that support operators to make certain decisions that affect the asset availability.

The intention of this book is to mention the most important tools that are applied to Asset Reliability and it is not intended to detail them, but to emphasize important aspects that are often not taken into consideration.

- **Process Reliability**

   The processes related to Operational Reliability can negatively affect it if they are not reliable within themselves.

   These processes can be related to production, administrative and managerial processes.

   The way to determine if a process is not reliable, is by analyzing the Key Performance Indicators that have been selected to control each process step.

   There are different ways to analyze a process and determine the gaps that prevent it from functioning properly.

   One of them is to analyze the indicators related to each one of the process steps (and the steps of the sub processes if they exist) to discover the deviations that prevent a good performance, plan and execute the necessary improvement and measure through Key Performance Indicators, if the proposed improvements generated the expected outcome.

   Another methodology is rather statistical and uses Lean - Six Sigma tools to define the deviations by statistical methods, to propose improvements and after implementing them, it also proposes to measure the effectiveness of the recommended actions. An approach proposed by Paul Barringer is available in his publication: Process Reliability and Six Sigma. In his proposal Six Sigma methods are used and complemented with Weibull analysis.

Regardless of how the process is analyzed, either manually or using statistical tools, the key is to ensure that the recommended actions to recover the reliability of the process are implemented and to evaluate their effects.

- **The Problem-solving Management Process**

The Problem-Solving Management Process interacts intensely with the Maintenance and Project Execution processes. It feeds from the data of both processes and generates Maintenance Strategies, Changes in assets to improve Reliability and provides feedback to improve these processes and their documentation.

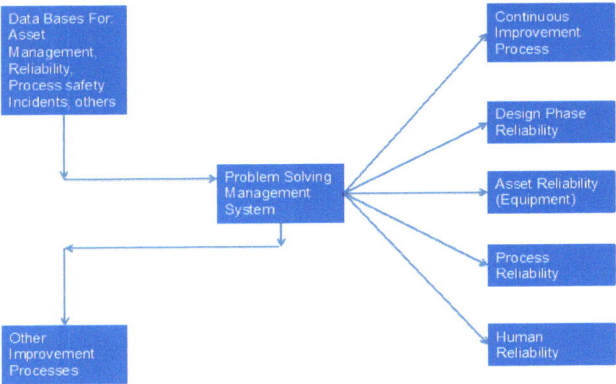

Figure 5: Management Process Reliability and Integrity

The processes that relate to the Reliability and Integrity Management Process are fed from the databases of the asset management system (SAP, Maximo, JD Edwards, etc.), the reliability management database system, the safety and environment incident database, and other databases that contribute to capture problems or situations that could affect the operational reliability of the facilities.

The data is entered into the Problem-Solving Management System. This management system contains a database that is managed by the work team of the industrial plant or facility to which they are associated. The members of this working group are the individuals who "naturally" interact during the day-to-day production and whose responsibility is shared with respect to maintaining a high asset availability. That is why they are called the Natural

Work Team. The natural work team is usually comprised by the representatives of operations (Day Supervisor, Instrument Panel Manager), Maintenance (Maintenance Supervisor, Planner), Engineering (Reliability Engineer, Process Engineer, Static Equipment Inspector, Rotating Equipment Engineer), occasionally people from safety, health and environment.

Normally, the principal members of the Natural Work Team are the Operations Supervisor, the Reliability Engineer, the Process Engineer, the Maintenance Supervisor and the Safety, Health and Environment Representative.

This database contains a list of problems or situations (Threats) that are or could be affecting the performance of the industrial plant and directly linked to the assets or processes that support the management of the plant.

This database also serves to rank these Threats and, depending on their relative importance and their nature, associate them with the different means that will serve to give an adequate response. The hierarchy is carried out applying the risk-based hierarchy methodology. The most critical threat is one that represents the greatest risk to the operational reliability of the asset. The ranking of the activities in the list must be done using the risk matrix of the industrial plant or the company. For this, the question that must be answered is: What is the risk for operational reliability if this threat or this item is not resolved? The higher the risk, the higher the priority of the item on the list. The risk matrix must not have more than 5 levels of consequence and 5 levels of probability (or frequency). A 5 x 5 matrix is sufficiently detailed to provide a good result in the hierarchy.

The database must also collect the decisions agreed by the natural work team and dates set for the execution and for future monitoring. A good tool that supports the management of this database is one that handles work orders (WO) in the asset management system (SAP, MAXIMO, etc.). Using the different status for the WO that are available, the execution of these activities can be managed.

In the cases where the above-mentioned system is not available, one can be created, for example through the application of an Excel spreadsheet or another data management tool, where the risk matrix and the logic to assign priority to the items are associated to the list.

If the threat is simple in nature, it only requires a simple analysis and the required action can be scheduled for immediate execution and removed from the list.

When the item has a greater complexity and a basic analysis does not allow to make accurate decisions, it should be resorted to more complete problem-solving methodologies such as the linear regression method (Kepner-Tregoe) or the Root Cause Analysis (RCA) methodology. In the context of this book, only the latter will be addressed.

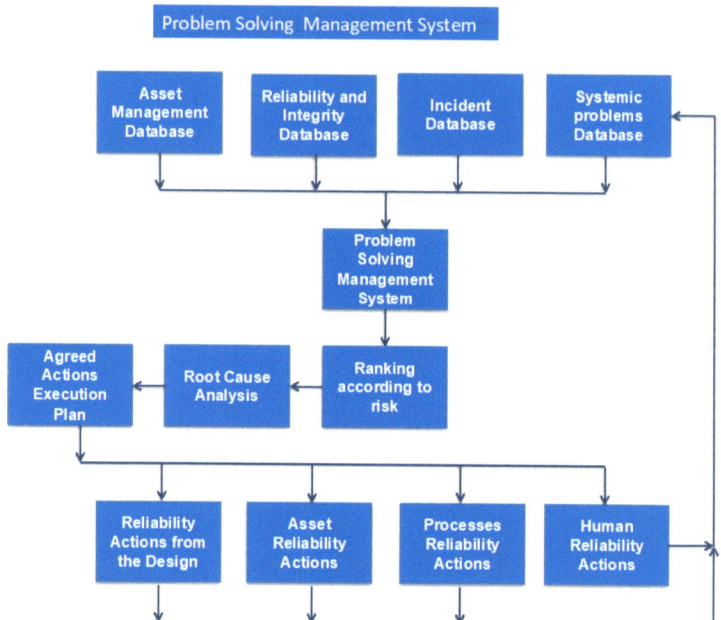

Figure 6: The Problem Solving by Management System

The Root Cause Analysis is a problem-solving methodology that in a very structured and is rigorous way defines what the primary cause is, or primary causes are that originate or originated the deviation or the occurrence of an unwanted event. In essence, this methodology seeks to answer why an event occurred within the chain of activities that contributed to the generation of a failure or deviation from a rule. While responding, it goes deeper in the search of the root cause or root causes until reaching them.

Although this is a methodology that is applied reactively, it can be used to analyze problems or potential situations and then define actions to prevent its occurrence (Analysis of Mode and Effect of Failure and Criticality, [FMECA]).

When initiating the resolution of the problems with the highest priority, improvements in the performance of the assets begin to be generated and in addition, time is freed that would normally be occupied in the attention of a recurrent problem, with the consequent reduction in maintenance costs and the associated opportunity losses. The time released is used to remove more waste from the processes associated with production.

The continuous improvement creates greater opportunity to generate more added value and more chances to attack the following causes of waste.

The root causes of the problems that affect the operational reliability of the assets can be related to its four aspects: Human Reliability, Design Reliability, Reliability of the Assets, and Reliability of the Processes.

Each one of the aspects has a range of proposals and methodologies that in a structured and systematic way supports the resolution of root causes.

Having come to define the root causes of a problem or situation allows generating a list of actions that will be aimed at eliminating those root causes. It is essential to ensure that the proposed solutions are effective and efficient, so that the repetition is not probable or that mitigation measures have been defined to avoid an inconvenient frequency, until a permanent solution is achieved.

The proposed actions to eliminate the root causes should be included in an implementation plan that defines for each action an entity responsible for its execution, which resources should be made available and the date of delivery of the proposed solution. The organization or individual responsible for monitoring the execution of these actions must report on the progress and on any obstacle that is obstructing their progress, so that, if necessary, this situation can be raised to the relevant decision levels that seek the means to remove these obstacles. With the above I want to emphasize that it is imperative to put as much effort as possible so that all agreed actions for the resolution of a problem or deviation are met.

It is also necessary to confirm that the action proposed and put into practice demonstrates that it is really providing the expected result. The definition of key performance variables supports this activity.

All the information related to the Problem-Solving Management System must be maintained in a database so that it serves for later systemic analysis or to avoid having to repeat an analysis, and, something completed previously can be used.

7. The Project Management Process

In the context of this book, we are dealing with the Project Management Process, from the point of view of the effect that this process could have on the operational reliability of the associated asset.

The Project Management process must be accompanied by a standard that dictates the guidelines so that the reliability, maintainability and operability aspects of a facility are considered throughout the life-cycle of the asset. For this purpose, criteria related to the aforementioned aspects must be applied, throughout the process and its execution phases.

There are methodologies that, well applied in each of the execution phases, will support the project being delivered with a high probability of showing high availability if it is operated within the context for which it was designed and built.

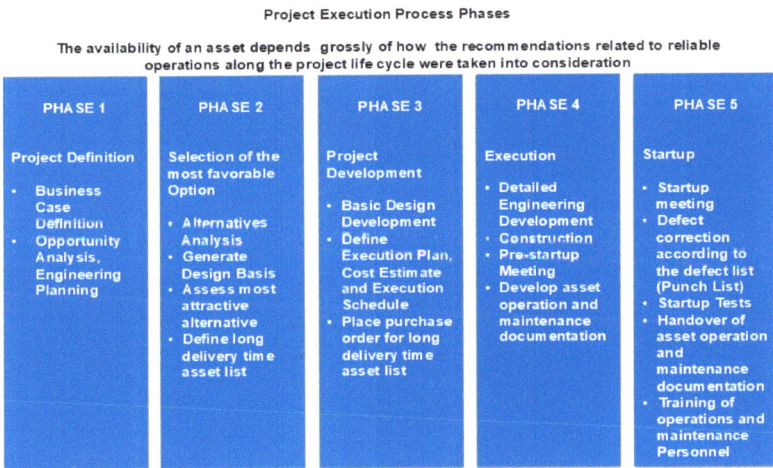

Figure 7: The Project Execution Process

When a project is developed, the factors of cost and execution time come into competition, with the factors of quality, reliability, maintainability and operability.

The engineer or the project manager will be interested in delivering a project at a lower cost and shorter execution time.

The owner of the asset must demand that the facility delivered to him be designed and constructed to operate reliably within the operating window that was established. Likewise, it must be delivered after verifying the proper functioning of the mechanical equipment, the control and safety systems, and having ensured the existence of all the documents required for an operation and maintenance of the equipment that make up the asset. It is equally important to have an adequate level of start-up and operation parts for the first two years, as well as having a team of well-trained operators and competent maintenance personnel.

During the development of the project, competent and committed operators and maintenance personnel must be integrated early, so that they can contribute with their experience and ensure that operability and maintainability aspects are considered during the design of the facility. Likewise, these must familiarize themselves with the new industrial plant and support the preparation of procedures for the proper operation of each of the equipment and systems. The maintainers will have to develop repair procedures of the equipment, as well as generate the maintenance strategies and the bills of materials and spare parts for the equipment starting with those critical ones.

As already mentioned, in each phase of the execution of a project, activities that contribute to greater reliability, maintainability and operability of an asset can be included. During the development of this section, those tools that I consider that most contribute to obtaining a better performance of the facility throughout its life-cycle will be mentioned, the best known, with the awareness that there are many others that contribute to excellent results.

During Phase 1, the most important step is the definition of the different technological options that can be selected to obtain the economic incentive that is being pursued with the implementation of the project.

The availability level of the industrial plant or facility is an aspect that must be established in phase 1, because it is essential to determine the profitability of the available technological options to make the proposed project feasible. For this, it is necessary to define the configuration of the equipment that intervenes in the production process. Depending on the need for redundancy that must be installed to meet a certain availability, the cost of the project will vary, and the profitability will be affected. One way to get to the most convenient configuration is to simulate availability through a Reliability-Availability-Maintainability (RAM) analysis. To do this, a block diagram of the equipment configuration is prepared, and reliability values are assigned to each of the blocks representing the equipment. These values are selected from available databases for the different types of plant or facility or are collected from the reliability databases of similar facilities. Care should be taken to ensure that these values are supported by a maintenance infrastructure like what is expected in the selected location to locate the proposed plant. Then the reliability of the set is determined using the RAM analysis. The resulting reliability is a measure proportional to availability. With the information of typical values of typical repair times it is possible to define the availability (which depends on the typical average times of operation and repair). There are programs that facilitate RAM modeling and with this, the simulation of different options is greatly facilitated, and this simplifies decision-making to reach the most convenient configuration of the plant.

During the first phase of the project, one of the most important tasks must be the definition of the useful life of the proposed asset. With this, it is possible to estimate the investment in reliability and the maintenance effort required to operate reliably during the life-cycle.

A successful project is one that provides the highest level of return on investment throughout its life-cycle. It is therefore important to ensure that the life of the project is clearly defined, because for projects of short existence, a high initial cost is not justified and vice versa, for a plant or facility with a long-life expectancy, it is necessary to increase the initial cost so that it has greater availability.

Applying the methodology of life-cycle cost analysis facilitates the determination of the initial investment cost and the cost of operation throughout the life-cycle. The most attractive option is selected, considering, in addition, any aspect of safety and environment that may have an influence on the selected option, as well as its potential to increase production with a sensible investment.

During Phase 2 of the project, the different alternatives are analyzed, and the most attractive option is selected from the point of view of the life cycle cost.

Depending on the level of complexity of the proposed alternatives and the investment, the proposed methodologies will be applied with greater or lesser rigor.

After having selected the most attractive alternative, a reliability analysis is initiated to detect areas where the design must be assessed in greater depth so that the reliability goal is reached. It is useful to apply the RAM methodology to refine the reliability values depending on the decisions taken in this phase of development of the basic design.

It is important to define in this Phase, what will be the performance and safety tests (control and interlock), reliability and integrity that will be executed during the pre-start and start up, so that they are prepared and located within the construction plans and in the start-up phase.

In Phase 3 it is necessary to emphasize that the design intrinsically has reliability criteria that cannot be ignored to guarantee the availability of the industrial plant and contribute to the safety of the processes. To ensure that these criteria of reliability in the design have not been left aside, it is customary to revise the designs so that there is a positive response to questions such as:

- ✓ Will the proposed design meet the expectations of the project from the point of view of reliability, integrity and safety of the processes?

- ✓ Will the design guarantee an unimpeded construction phase?

- ✓ Will the design guarantee a clear operation without surprises?

- ✓ Can the design guarantee good access and ease when executing equipment maintenance?

All these questions made throughout the period of Phase 3 and especially in the period of definition of the basic design, may generate design solutions that respond adequately to the concerns that arise when analyzing them.

It is highly recommended to review the design during this Phase, which is executed by independent professionals in interaction with the project design team.

Because of this review, adjustments may have to be made that increase the operational reliability of the project and that require a life cycle cost analysis that proves the convenience of the change.

Phase 4 is the one where the project is executed. This begins with the completion of the detailed engineering. The process flow diagrams and the process and instrumentation diagrams are ready, and the detailed engineering must be completed and the equipment with a long delivery time must be already ordered during Phase 3. The specification sheets and the construction drawings must be revised to ensure that they are in accordance with the design intent and the procurement of materials and equipment is initiated.

As stated before, it is absolutely necessary and convenient to successfully complete a project, to include operations and maintenance representatives as part of the project team from the beginning.

The operations representatives will be responsible during this phase to prepare the operation procedures of the facility. They must also generate the training material and impart it according to the project execution plan, so that when it is delivered, the operators are able to operate the facility in a reliable and efficient manner. This way, entering with the right foot to operate the facility is guaranteed, in accordance with the aspects of operational reliability, especially considering the human reliability.

The maintenance personnel assigned to the project team will have as tasks during this phase, to ensure that the hierarchy of equipment of the new facility is defined, to generate the master data of equipment and materials of the industrial plant, as well as the list of materials associated with each one of the equipment. They must also produce the maintenance strategies for the facility and the specific lists of materials associated with the equipment. Based on the equipment hierarchy they should coordinate the meetings to determine their criticality. These meetings must be attended by the members of the team responsible for the operation and custody of the new facility. At least, the representatives of operations, maintenance, reliability engineering and process engineering, the person in charge of inspection and, if possible, a safety representative should be part of this task. This is an exercise that requires the active participation of these individuals. Once the criticality has been defined, the maintenance strategies dedicated to each of the teams and their corresponding asset care plans will be generated.

The maintenance representative will be the individual responsible for preparing the equipment files and together with the project manager ensure that the engineering company and the equipment manufacturers deliver all the required information that is necessary for operational reliability management.

Another important task that refers to the maintenance of the equipment is related to the procurement of any special tool, equipment or service necessary for the management, as well as its custody.

The representatives of maintenance and operations must agree on means for the process of safeguarding these means of maintenance support to be well kept over time.

The maintenance representative of the project is also committed to the creation of training material for the different occupations that will participate in asset care activities and will place special emphasis on those teams of high criticality that require special attention due to their complexity.

The maintenance representative should prepare all the condition monitoring systems, develop preventive maintenance plans and ensure that all data and plans are uploaded into the computerized maintenance management system.

It is important that maintenance and operations representatives initiate the development of the maintenance work management process based on the risk-based work management methodology. Even if there is a system of criticality of the equipment and a methodology of assigning priorities to pending work depending on the type of maintenance, applying the risk-based work selection methodology allows managing work orders avoiding performing low priority work allowing to execute WO with higher priority. This methodology allows managing the backlog more efficiently.

The walk-through of the plant to detect defects that will prevent proper operation and maintenance and facility failures allows timely correction of deviations before delivery of the project.

At the Pre-Start Meeting, how the deviations detected in the walk-through of the plant will be corrected must be agreed upon. There is also an analysis of compliance with the control/quality assurance programs and the proposed changes. Especially, if they could have generated a negative effect on the operational reliability must be analyzed. Another aspect that is reviewed is that of safety and the safety systems that must be inspected to have them ready at the time of startup are agreed upon.

The availability and quality of the products (catalysts, reagents, additives, start-up parts, tools and special equipment) that are required for the proper operation of the plant and its proper storage should be confirmed. Likewise, it must be ensured that the process and maintenance information (procedures, guides, drawings, plans, etc.) is available in the control rooms and maintenance offices. The computerized maintenance management system must be accessible already during the construction phase and the training in its modules and use must be given to all maintenance people and those of operations that have common tasks with maintainers.

8. The Importance of the Data, Data Quality and Information

For the purpose of this book, the Data is a descriptor element of a condition associated with a physical entity that may suffer variations over time. A collection of data can be analyzed and result in information that contributes to the decision making related to this entity.

With this we want to affirm that a set of data can generate valuable information that supports the decision making regarding a condition or situation that we want to modify to obtain superior results.

There are multiple categories of data that are related to the management of companies and more specifically to the operational reliability that we can differentiate as follows:

- Asset descriptors, which are commonly referred to as **Master Data**. These are elements that detail to individual, an asset or geographical locations in their composition and physical or intangible attributes and are specific to each asset, individual, geographical location that the company manages.
  This master data is essential in a modern company, because it provides support in the management of its assets.
- Descriptors of performance behavior. These can be measures of physical conditions that are related to the operation of an asset. Others may be key performance measures that are developed to evaluate the behavior of a process or a

business function, etc. In general, this data is measured and compared against a pre-established pattern or against a key performance measure that is set as a limit.

The management of business data has become increasingly important and is undoubtedly a key activity of any company.

For purposes of operational reliability, the two types of data described above are of particular importance. The master data ensures a correct administration of the assets from the point of view of their relative importance for the generation of value, for the maintenance management and the administration of spare parts that are required to have available in the storage of materials of the company.

The master data may refer to the equipment of an industrial plant, to the materials and supplies related to its operation, to the preservation activities required to ensure a defined plant availability, and they support the hierarchical structure of the assets that intervene in the operation of the facility.

The reliable management of this data serves many purposes, such as serving the finance organization to maintain an adequate accounting of the physical assets available to the company, allowing it to track the operating costs related to each of them. These assets allow associating expenses with organizational entities that operate the equipment, etc.

The operations organizations have a powerful tool in the master data regarding the association of operational costs, assurance of spare parts and parts for these assets, depending on the relative importance (criticality) they have for the productive processes of the plant, etc.

As the master data supports different business processes within the company and in a world where information technology is gaining increasing importance, it is important to generate standards that regulate the formatting of the master data, the conventions and associated attributes, so that these are correctly integrated into the management systems of the company.

The master data usually has documentation associated and information related to it. This information, in the event of operational reliability, can be drawings, drawing arrangements, lists of materials and spare parts, maintenance manuals supplied by the manufacturer, maintenance strategies associated with the asset, among others. This information associated with a master data is called Master File.

The data describing behavior or performance are transactional and reflect values or periodic measurements of physical parameters or observations.

This data can be related to the master data and included in a database that is associated with the measured or observed asset.

Although it is true that the data of assets that intervene in a productive process are largely related to their performance, transactional data that has to do with the costs that are caused by the maintenance and using inputs related to its operation is also filed. The data of work orders for example, are documents that after their closure, must be saved for further analysis and eventually generate an improvement of the processes or behavior of the asset.

Data from the different databases that are related to the master data can be analyzed in a singular way and related to a specific asset or can be analyzed in a systemic way to generate improvements to a diversity of cases that without the existence of systematically stored data and in an organized way, it would not be possible.

The systemic analysis of data from diverse databases in a cross-way is a very powerful tool to detect opportunities for continuous improvement.

The analysis of data collected by a vibration variable monitoring system in conjunction with the temperature data obtained in the same way, can help to detect defects of a machine that is approaching an unacceptable level of behavior that could generate a serious failure of an asset, in maintenance language focused on reliability, approaching Point F of the PF diagram. This is just one example of data analysis applications that can be performed to improve an asset's performance. An analysis of the data of two databases analyzed concurrently can generate valuable information that greatly facilitates the making of an accurate decision.

The set of work orders that contain previous data of the asset, is an example of what is called Historical File. It is an important compilation of data that facilitates failure analysis and in general support continuous improvement. It is also a source of information that meets the reporting needs that regulatory entities may demand, as is the case of regulations related to equipment integrity or the environment.

In any WO, repair data (which may be minor or major) is stored with dates associated with them, costs of the maintenance performed, among others.

The portion of the WO that must be completed in a reliable and professional manner is the one that refers to the conditions in which the asset was found at the beginning of the intervention and whiech are the associated failure codes that are considered descriptors in the initial diagnosis.

Afterwards, a written report must be made that includes important details and measurements, how the asset was left after being intervened to restore its reliability condition.

It is essential to leave a detailed repair report.

WO's include cost data associated with them that are an important part of management measurement and should be charged to the WO either automatically if an asset management system is available that allows it, or manually.

The question as to why to be so emphatic with the registration of all this data has a simple answer: It is not possible to benefit from good asset management without good INFORMATION.

The data included in a well-conducted WO will serve to multiple activities related to the care of the assets:
- **Analysis** of equipment failures.
- Definition of availability levels of spare parts.
- **Analysis** of interchangeability of spare parts.
- Equipment Standardization.
- Analysis of asset care costs.
- Data source for the presentation of management indicators.
- Generation of information to demonstrate compliance with safety and environmental regulations.

The list is much longer.

With this example, what we want to prove is the importance of registering, maintaining, using in a professional manner and safeguarding the data associated with the management of operational reliability. With high quality data and the required amount, you can generate information that will allow you to make sound decisions that will contribute to an excellent management and to the continuous improvement of the processes, and thus to a greater profitability of the company.

Every company must have a system that guarantees the appropriate management of the data associated with its management. This system must guarantee the selection, storage, protection and proper use of the data it requires for optimal operation.

This system is called Data Governance and is becoming increasingly important in the technified and globalized world of today.

The **Data Governance** is a method or system that using personal expertise, data management processes and appropriate support tools, makes available to the company a data bank that guarantees that the information generated based on the extracted data of it, be reliable and dependable.

The Data Governance allows a more transparent management of the company, facilitates the administration of management reports internally and to auditors and regulatory bodies. It provides a large competitive advantage for the company that can prove to have an effective Data Governance.

The greatest benefit of an effective data governance is the availability of reliable data that supports continuous improvement.

9. Measurement of performance in Processes

Measurement is an activity that dates back thousands of years. It arose from the need to determine the distances that were necessary to travel to ensure a hunting place with a good promise of abundance of prey, for example. Then it was perfected with the awareness of the concept of value that the human being has developed during its evolution.

Measuring is comparing one magnitude with another that is standard and that represents a unit of comparison or measurement.

On the other hand, a process is a succession of steps or activities that must be executed to achieve a result. The notion of process in our modern life is practically part of everyday life and is used in different fields of human endeavor.

Successful companies in this highly technified and competitive world in which we live, have developed, documented and implemented their business processes generally supported by a technological solution that allows them to define and establish key performance measures, generating a series of data that can later be analyzed.

Each step or activity of a process has inputs that are transformed within its scope and products that result from this activity or step. That is to say, that each step or activity of a main process is in turn a subprocess and as such can be subject to measurements.

Everything that is measurable generates control data that can be associated with quantity, quality, physical properties, chemical composition, as well as intangible properties such as, for example, the level of customer acceptance, among others.

There are also reference data that can be used to determine if there is a gap between the data measured in the evaluated process and the reference. Depending on the economic attractiveness of closing this gap, the necessary actions would be taken to approach the reference data or even overcome it.

You can also use the different data related to a process to perform an analysis of these and correlate the results to achieve an improvement in the process that is being evaluated. There are several methodologies for continuous improvement offered by quality management to carry out these analyzes and bring a process to higher values efficiency and effectiveness, according to a cost-benefit analysis that justifies the effort to optimize.

The tools for continuous improvement have undergone changes in their designation and have combined and/or modified methodologies that have led to new offers of solutions that are essentially based on the same basic principles. An evolution that is very much in vogue today is the application of the principles of Lean-Six Sigma.

An excellent combination for the continuous improvement of the processes is the one that is obtained when analyzing data of a plant operation database for example and the deviations that are suffering the process parameters are defined (key variables of yield). Then a Root Cause Analysis is carried out to determine the reason for the deviations, an improvement plan is formulated, and the agreed actions are taken to solve their causes.

A very illustrative example would be the measurement of the sulfur content of a gas oil stream from a hydro-treater in an oil refining complex. It is being noticed that the sulfur content is reaching the permissible limit and it is necessary to correct this anomaly because it could affect the placement of a lot for sale. When analyzing the sulfur concentration data on their own, no conclusion is reached about the origin of the deviation. However, when performing a root cause analysis and observing the entire process, it can be concluded that the activity of the reactor catalyst has been diminished by the fact that contaminants have affected it. The failure of an exchanger could have been caused by an accelerated corrosion, which in turn had its origin in an inadequate selection of materials. This wrong selection could have been caused by a very optimistic specification of the sulfur content of the current that must be cooled in said equipment. The optimistic specification of the sulfur content could be due to the short experience of the engineer who was designated for the task of producing said specification.

Continuing with the analysis, it could be detected that the rupture of a heat exchanger is the source of entrance of the pollutants that affected the activity of the catalyst in the reactor. The solution would then be to replace the heat exchanger with one that is manufactured with the appropriate materials and would not be the action that is often taken as the most immediate would be to repair the equipment. This is to illustrate; how important it is to review all the processes involved in the operation of a facility.

It should not be forgotten that the operational reliability of a facility depends on various business processes that interact and could influence each other. The proper analysis of the processes and their interactions is therefore vital.

The continuous improvement of business processes is a key contributor to achieve optimal operational reliability.

10. Epilogue

Dear reader,

I hope the effort that was put into publishing this book will be of your liking, and that it helps those who are beginning in this field to understand what Operational Reliability is. I also hope it is a source of strengthening and consolidation of knowledge for those who are already on the path of this discipline, and finally I wish that those versed in the topic find this work useful in some extent.

The language was kept simple because I am convinced that it is the best way to convince the reader of the usefulness of the application of these concepts and methodologies. We did not delve into applications of the related techniques because there are many publications that address these aspects and thus make this effort unnecessary.

In the near future, I intend to present my experiences regarding the implementation of this process and emphasize on ways to facilitate the transition to a new way of managing assets so that they present optimal performance and to achieve improved numbers of Facility Availability or of Overall Equipment Effectiveness (OEE). All this supported by better efficiency in the application of this process.

I also want to emphasize that continuous improvement of the processes, is a fundamental part for the excellent performance of any business activity and as such it should be an integral part of the applications of all the business processes that any organization has.

I deeply appreciate your attention and I invite you to send your suggestions so that this work maybe be improved.

Alberto Windmüller

11. Bibliographic References

1. Documentos relacionados al Proyecto de Transformación de Mantenimiento de la Refinería Cardon, MARAVEN S.A., Filial de Petróleos de Venezuela, 1992-1995

2. Material del Curso Cost Benefit Analisis, de la empresa The Woodhouse Partnership, Londres, Reino Unido, 1994

3. RCM 2, John Moubray, Second Edition, Industrial Press

4. API 580 RP Risk Based Inspection, American Petroleum Association

5. Apollo Root Cause Analysis, Dean L. Gano, Apollonian Publications, Yakima, Washington

6. A culture of Rapid Improvement, Raymond C. Floyd, CRC Press

7. The Discipline of Market Leaders, Michael Treacy & Fred Wiersma, Basic Books, a member of Perseus Books, New York

8. Ten Steps to Quality Data and Trusted Information™, MIT Information Quality Industry Symposium, July 15-17, 2009

9. TOPICOS, Conferencia Plenaria de Mantenimiento y Confiabilidad, Petroleos de Venezuela 2001, Caracas

10. 7 phases to build reliability into a capital Project, Plant Services, http://www.plantservices.com

www.ingramcontent.com/pod-product-compliance
Lightning Source LLC
Chambersburg PA
CBHW040227220526
45473CB00001B/145